For Dane who always figures out the answers...

Science Advisor
ROBERT L. WILLIAMS, KENNEDY SPACE CENTER

Owlbop

NO PART OF THIS PUBLICATION MAY BE REPRODUCED, STORED IN A RETRIEVAL SYSTEM, OR TRANSMITTED IN ANY FORM OR BY ANY MEANS, ELECTRONIC, MECHANICAL, PHOTOCOPYING, RECORDING, OR OTHERWISE, WITHOUT THE WRITTEN PERMISSION OF THE PUBLISHER, LINDSEYCLARISSA@GMAIL.COM

Silly Scientists Take A Peeky At The SOLAR SYSTEM!

Written by LINDSEY CRAIG

Illustrated by
MARIANELLA AGUIRRE   YING HUI TAN

We're in a Solar System.
Yes we spin around the Sun!
Our planet is the Earth,
so we're Earthlings everyone!

That's NEAT-O!

4.6 billion years ago
things started spinning round
when a space cloud
CRASHED
on its fanny like a clown!

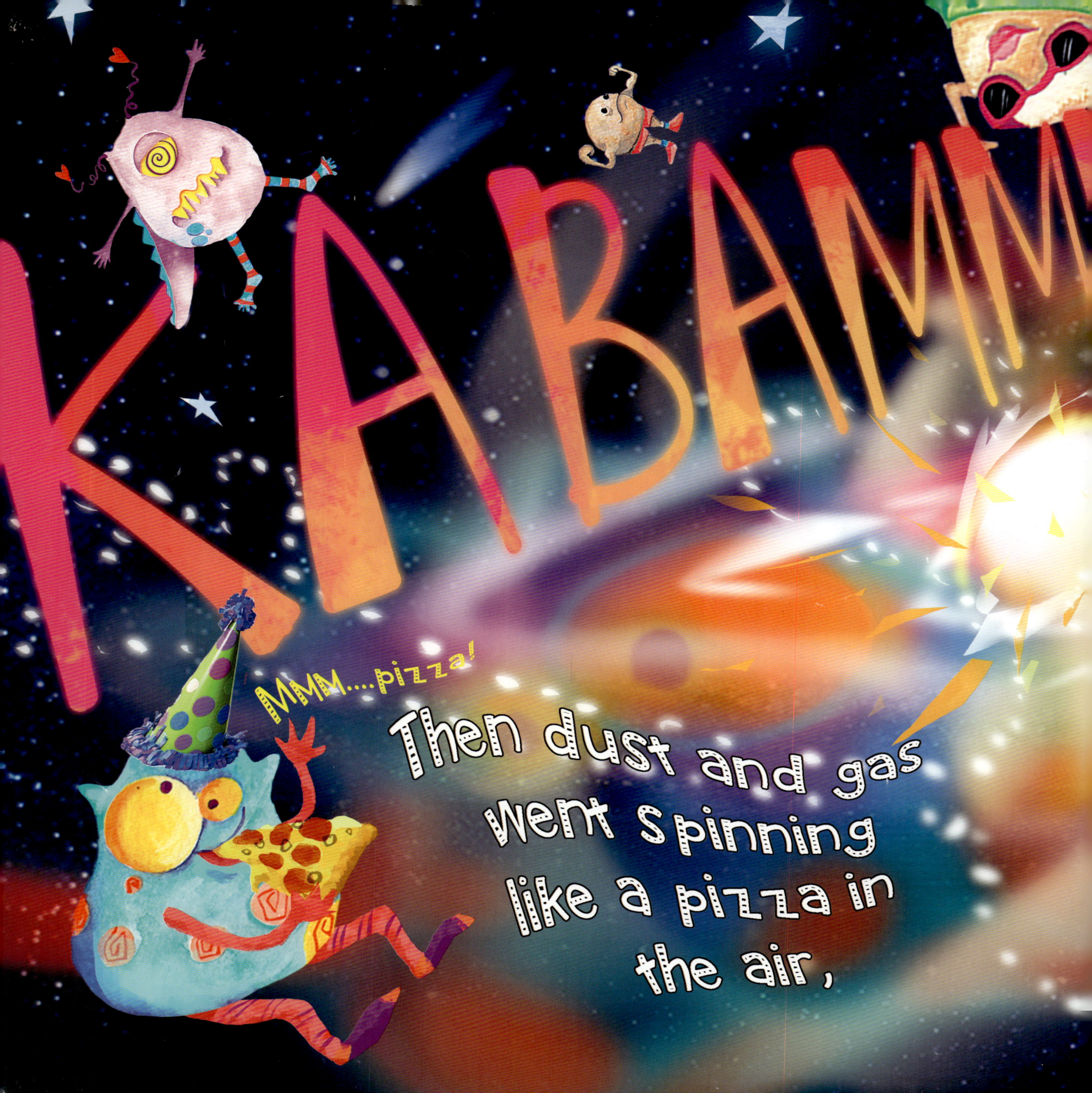

whoochie!

Now **Mercury** is the fastest! Watch it ZOOM around the Sun. With frigid nights and days hot-HOT! it scorches little buns!

And Earth's
our water planet,
so lovely, wet and warm,
with prairies, mountains,
oceans and life of
every form.

Mi Mamá Tierra!

Number six is super Saturn.
It has rings like colored glass.
It's a whopper of a planet,
but it's mostly fluffy gas.

Euwwwww!

Uranus is an ice giant that has a wonky ride. Its seasons last forever since it orbits on its side.

"Dude! No hurling on the planet."

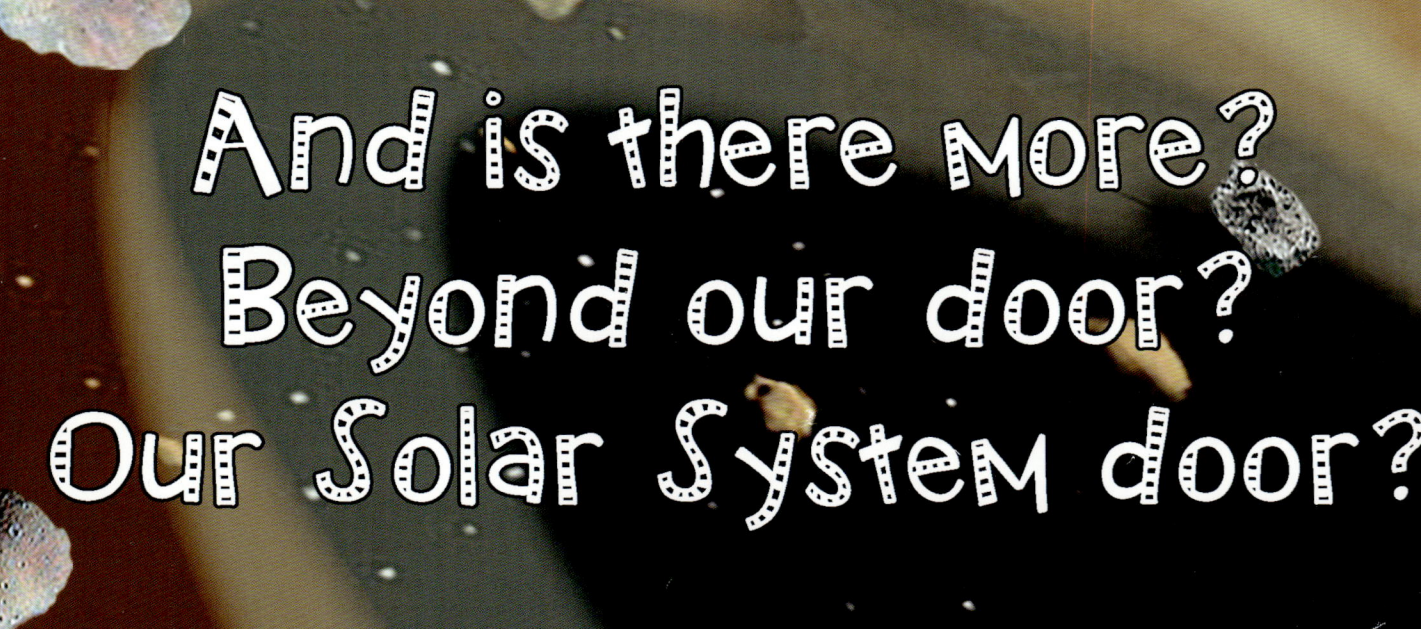

And is there more?
Beyond our door?
Our Solar System door?

Whoochie!

THE SOLAR SYSTEM is made up of the Sun (a star!) and everything that is in its whirlpool-like gravitational pull, including eight planets, five dwarf planets, hundreds of moons, asteroids and comets, the Kuiper Belt (rhymes with sniper snelt), the Oort Cloud and every bit of gas and dust in between. Our Solar System is approximately two light years across, or about 12 trillion miles (12,000,000,000,000) or 18 trillion kilometers.

A NEBULA or Nebulae (plural) is the thickest part of Giant Molecular Clouds. These space clouds are gargantuan, like enormous dust bunnies—at least 100,000 times larger than our sun. When parts of these clouds collapse, they form nebulae or knots of thick gas and dust. After millions of years, a nebula can collapse further and further (due to gravity) spinning at unimaginable speeds, getting hotter and hotter, until finally bursting into a star. Dark nebulae are often called star nurseries because within them often hundreds and even thousands of stars are born.

PLANET MERCURY blasts around the Sun at over 100,000 mph (161,000 kph). That's much, MUCH faster than a speeding bullet! And though Mercury zooms around the Sun, it rotates very slowly on it's axis, making a day on Mercury much longer than its year. A Mercury day is equal to 176 Earth days, while a year on Mercury is only 88 Earth days! Mercury is hot, too! The temperature during the day can reach 870° F (450° C), then drop 1,000 degrees at night to an incredible -240° F (-170 °C). In the picture at right, that little black dot is Mercury as it whizzes past the Sun.

PLANET VENUS is the hottest of all the planets in our solar system, roasting at 870° F (466° C). That's hot enough to melt lead! The heat of the planet is due to the Greenhouse Effect; this is created by thick CO_2 gas clouds that surround Venus like a wooly sweater keeping in the Sun's heat. Venus's atmosphere (air) is so heavy that just standing on the surface would crush you AND your spaceship into jelly!

PLANET EARTH is the only planet (of which we know) in our vast Universe that supports life. Scientists believe life formed some 3.8 billion years ago as primitive single-celled organisms. At the time of Earth's formation, asteroids bombarded our planet. These asteroids brought water to Earth, and one giant asteroid knocked a chunk off of our planet. This chunk became our moon! Earth is indeed a water planet with more than 70% of its surface covered by water. Surrounding us is Earth's marvelous atmosphere called the troposphere. Though this layer is only 4 -12 miles (7-20 km) wide, it contains ALL the air we breathe and All of our weather.

PLANET MARS is half the size of our Earth, yet it has truly giant mountains and canyons! Olympus Mon, the tallest mountain on Mars, towers 14 miles high; that's like stacking, end to end, 80 Eiffel Towers or 8 Golden Gate Bridges! And, the Martian canyon, Valles Marineris, would stretch across the entire United States. Mars is called the red planet because of the red iron ore soil that covers its surface. Violent Martian wind storms often kick up this fine dust creating an orange-red cloud around the planet. And though Mars appears red, it has sunsets that are blue.

Images: NASA